I0191253

Not to be taken into Action or Front Line Trenches.

THE TACTICAL EMPLOYMENT OF LEWIS GUNS.

ISSUED BY THE GENERAL STAFF.

This publication cancels the instructions relating to Lewis Guns in S.S. 106 (March, 1916), and S.S. 122 (September, 1916); and includes S.S. 194 (October, 1916).

January, 1918.

PRINTED IN FRANCE BY ARMY PRINTING AND STATIONERY SERVICES.

Published by the
The Naval & Military Press
in association with the Royal Armouries

Unit 10 Ridgewood Industrial Park,
Uckfield, East Sussex, TN22 5QE
Tel: +44 (0) 1825 749494
Fax: +44 (0) 1825 765701

MILITARY HISTORY AT YOUR FINGERTIPS
www.naval-military-press.com

ONLINE GENEALOGY RESEARCH
www.military-genealogy.com

ONLINE MILITARY CARTOGRAPHY
www.militarymaproom.com

ROYAL
ARMOURIES

The Library & Archives Department at the
Royal Armouries Museum, Leeds, specialises
in the history and development of armour
and weapons from earliest times to the
present day. Material relating to the
development of artillery and modern
fortifications is held at the Royal
Armouries Museum, Fort Nelson.

For further information contact:
Royal Armouries Museum, Library, Armouries Drive,
Leeds, West Yorkshire LS10 1LT
Royal Armouries, Library, Fort Nelson, Down End Road, Fareham PO17 6AN

Or visit the Museum's website at
www.armouries.org.uk

CONTENTS.

SECTION   PAGE

I.—Characteristics of the Lewis Gun 3

II.—Duties of Officers 6

III.—Tactical Training of Lewis Gunners 8

IV.—Lewis Guns in the Offensive—
    A.—Trench to Trench Attack 13
    B.—In Open Warfare 14

V.—Lewis Guns in the Defence—
    A.—In Trench Warfare 18
    B.—In Open Warfare 22

VI.—Co-operation 24

VII.—The Employment of Lewis Guns against Aircraft ... 26

APPENDIX
I.—Firing the Lewis Gun from the Hip 29

II.—Ammunition 31

III.—Lewis Gun Drill 33

IV.—Instructions for wearing the equipment for carrying
    Lewis Gun Magazines 36

PRESS A—1/18—5404S—45,000.

THE TACTICAL EMPLOYMENT OF LEWIS GUNS.

I.—CHARACTERISTICS OF THE LEWIS GUN.

1. At the present time an infantry battalion has sixteen Lewis guns. Pioneer battalions have eight Lewis guns and Cyclist battalions six.

2. The Lewis gun shares with the Vickers gun the following characteristics :—

(a) Action only by means of fire. Fire power can greatly assist an attack, but it cannot by itself gain ground.

(b) Rapid production of great volume of fire. The fire power of a Vickers gun, and, for a short time, of a Lewis gun, equals that of about 20-25 good shots with rifles.[*]

The feature of the fire delivered is its close grouping. If the gun be laid correctly it will hit very hard, but a slight error in aiming or ranging will make it miss altogether. To get full value for the ammunition expended, both guns require a deep target. This can often be obtained by surprising an enemy in close formation or by siting the guns so that they can bring enfilade fire to bear.

(c) Small personnel required. Only two men need be with either gun in action. This characteristic is of fourfold importance :—

(i.) Economy of men in defence.

(ii.) Invisibility. A Vickers gun or Lewis gun in action occupies about 6 square feet.

(iii.) Invulnerability. The guns can fire with undiminished volume so long as two members of a team remain. In a rifle section every casualty diminishes the volume of fire.

(iv.) Rapid change of front with a minimum of movement.

(d) Liability to accidental cessation of fire. Both guns contain delicate mechanism, and, either through breakages or through dust and mud, may temporarily be put out of action.

---

[*] Any such comparison can necessarily only be a rough average, varying according to the circumstances. Against a line of skirmishers advancing from cover to cover in very broken country, probably three first class rifle shots would be more valuable than a machine gun. In defending a flat stretch of country, a bridge or narrow defile, a Vickers gun, or a Lewis gun for a short period, is equal to 80 or 100 rifles. On a narrow frontage, as in a village street, it is impossible to place a sufficient number of rifles to equal the fire of a Vickers or Lewis gun.

(e) Noise of firing. This sound is unmistakable, so that, once it has been heard. surprise is no longer possible.

3. The Lewis gun differs from the Vickers gun in the following characteristics :—

(a) The Lewis gun is not capable of sustained fire. The Vickers gun is cooled by water, and can continue firing for a long time, so long as its barrel is kept covered. The Lewis gun is air-cooled, and will get over-heated when about 700-800 rounds have been fired rapidly. After about 1,000 rounds it will probably stop, and will take half-an-hour to cool.* Its fire is, therefore, better applied in short bursts.

(b) The Vickers gun, with its Mark IV. tripod, has a fixed platform. The Lewis gun with bipod has not. The Lewis gun is therefore lighter and more mobile, but less accurate than the Vickers gun. The advantages of a fixed platform are :—

(i.) Reduction of personal errors in laying and firing.

(ii.) Accuracy in night firing.

(iii.) Possibility of indirect fire, and greater safety of overhead fire.

The Lewis gun with bipod mounting weighs 28 lbs. ; the Vickers gun weighs 39½ lbs., plus 48 lbs. tripod. With the Vickers, a condenser and additional water must be carried.

With the Vickers. one man carries the gun, another the the tripod, and a third the first ammunition supply. A single mán can carry the Lewis gun with bipod, ready loaded with a magazine of 47 rounds ; he can, therefore, come into action on the ground instantaneously, or, if necessary, can fire a short burst from the hip.

The mobility of a gun depends, however, to a great extent upon the mobility of its ammunition, and Lewis gun ammunition in magazines is heavier than Vickers gun ammunition in belts. When a convenient ammunition dump has been established, the mobility of a Lewis gun is practically as great as that of a rifle. But in case of a long or rapid advance in action, the gun will outstrip the bulk of its ammunition unless sufficient men are available to carry magazines forward.

(c) While both guns are liable to have accidental stoppages, the Lewis gun is the less reliable weapon.

The Lewis gun contains a large number of small springs and other delicate parts, which are subjected to a great strain during heavy firing, and are apt to break or get damaged.

---

* Individual Lewis guns vary very greatly in this. There are records of certain Lewis guns firing 2,000 rounds with little cessation : a new gun will usually heat much more quickly owing to greater friction. Experiments carried out with selected guns, well supplied with oil, show that the stopping point usually occurs after 800-1,000 rounds, in cases where the gun is at once reloaded after each magazine is finished. The first sign is often the premature explosion of a cartridge on touching the chamber.

While they can be quickly replaced by well-trained men, the gun must for a time stop firing. The Lewis gun is also more open; dust and mud can get into the works very easily, so that great care has to be exercised in crawling forward into a position. If the gun gets badly clogged the only thing to do is to take it to pieces and clean all parts.

The magazines of Lewis guns are also more fragile than the belts of Vickers guns, for a slight dint in a magazine will prevent it from fitting the gun or hinder the cartridges from passing out. They must therefore be carried with great care.

II.—DUTIES OF OFFICERS.

1. An Infantry Battalion has 16 Lewis guns, and should maintain 16 Lewis gun sections at all costs. One section in each platoon is trained in the use of the Lewis gun, and lives, fights and works with the other three sections. With an average strength of 36 O.R. in the platoon, the Lewis gun section will consist of 1 N.C.O. and 8 men.

2. Although this will be the normal organization, it may often be advisable in dealing with particular situations to allot a second Lewis gun to a platoon, or to withdraw temporarily one, two or more guns from platoons for special tactical employment under the orders of company or battalion commanders. Lewis gunners may also at times be withdrawn from their platoons for instructional duties.

3. A *Lewis gun officer* is attached to the headquarters of every battalion. His duties, under the battalion commander, comprise :—

(a) Advice to the battalion commander as to the allotment and employment of Lewis guns in attack and defence, choice of positions, ammunition supply and requirements, routes to be followed, use of limbered wagons to economize labour, special tasks, etc.

(b) Preparation of schemes in co-operation with the O.C. machine gun company or section, in order to ensure co-ordination between all concerned.

(c) Courses of instruction, and to arrange that the best value is obtained from the time available for special training.

(d) Test of all newly-arrived Lewis guns before they are placed in the line.

(e) Periodical inspection of all Lewis guns and examination of their teams.

(f) Inspection of any gun reported defective.

(g) The necessary arrangements with the Ordnance Department and with armourers. This includes the examination of broken or damaged parts before and after repair, and the decision as to whether such parts should be mended by an armourer or should be returned to the Ordnance Department.

(h) Advice to company commanders as to Lewis gun firing-places, shelters for guns and ammunition, and superintendence of their construction.

He should report to company commanders on the capabilities of men he has under instruction. If company commanders afford him the opportunity, he can impart much useful instruction at odd times, e.g., when guns are being cleaned.

(*i*) If the battalion is short of competent officers, and there is no pressing work in connection with Lewis guns, the Lewis gun officer may be temporarily attached to a company. But with 16 guns and 16 sections to be maintained in a state of efficiency, such occasions are likely to be rare.

4. The *company commander* is responsible to the battalion commander for seeing that his platoons have an adequate number of trained Lewis gunners, for the condition of the Lewis guns in his company, and for the training of their teams, and in general for their tactical employment in action. In particular, he must ensure co-operation between the Lewis guns of different platoons. In operations where the company Lewis gun limbered wagon can follow closely, he will provide a trained man to direct its movements.

He must remember that the Lewis gun is not a simple weapon. Its care and handling require knowledge which the Lewis gun officer has had special opportunities of gaining, so that even if the latter has had less experience of fighting than the company commander, his advice on Lewis guns should always receive full consideration.

5. The *platoon commander* commands the Lewis gun in action. At many stages of a fight he is the only officer who can affect the issue, so that he must have a thorough knowledge of the tactical uses of the weapon.

He must know enough about mechanism to judge whether his Lewis gunners are sufficiently trained, and must report any weakness to his company commander, in order that time may be allowed in the schedule of company training; it is useless for him to manœuvre his gun into the best tactical position if his gunners are not sufficiently trained to take advantage of the opportunity. He must also report at once any shortage in trained Lewis gunners. In allotting tasks to his platoon he must take into account that Lewis gunners have additional work in the way of cleaning or repairing guns and magazines and overhauling spare parts and ammunition.

III.—TACTICAL TRAINING OF LEWIS GUNNERS.

1. *General Principles.*—The Lewis gun section will normally carry out their tactical training with the other sections of the platoon under the platoon commander, who will normally command them in action.

But before a Lewis gun section can take part in platoon training it requires separate instruction in details of team work. While all the team, except the two actually on the gun, carry rifles and bayonets in action, and must be prepared to use them when required, they also have certain duties to perform with regard to Lewis guns, either as scouts, observers, or ammunition carriers, and must be so trained.

2. *Duties of the N.C.O. i/c of the Lewis Gun.*—The *sergeant* or *corporal of the Lewis gun section* leads the section into action, chooses the actual position for the gun, and gives the fire orders. He must be specially trained in fire control, in judging distance and in observing fire. He must be practised in quick decision, control of men and ability to make his meaning clear in the fewest possible words, or by some simple system of signalling. In an attack, when one platoon has been reinforced by others, he must reorganize his section at the earliest opportunity to prevent magazines being separated from the gun.

He must be thoroughly trained in the Lewis gun, and realize its difficulties and possibilities. In the event of the Lewis gun having a prolonged stoppage in action, he will at once reorganize the remainder of his team as a rifle section, while Nos. 1 and 2 try to remedy the stoppage.

3. *Special Training.*—The following points require emphasis in any special Lewis gun training :—

(a) Each man of the gun team must be trained to take any position in the team, and must be able to examine the gun in detail, to clean it without damaging it, to fire accurately at a given target, and to remedy stoppages without delay.

(b) No. 1 carries the gun into action, loaded with one magazine, and fires it. No. 2 acts as his assistant, lies beside him while firing, and carries spare parts and additional magazines (normally four). No. 1 and No. 2 are armed with revolvers, and must be trained in their use.

The remainder carry rifles and rifle ammunition.

No. 3 carries a supply of magazines and acts as the link between the gun and its ammunition supply. In action he will keep in close touch with Nos. 1 and 2, and bring forward to them magazines as required.

The remainder of the magazines, the number of which will depend upon the requirements of the operation, the state of the ground, and the distance to be traversed, will be divided among the rest of the team ; but there are advantages in having two men less heavily loaded than the rest, in order that they may move freely as scouts or runners. Four magazines (in addition to his rifle, ammunition, and ordinary equipment) are a fair load for one man, and eight per man should not be exceeded.

(c) All numbers must be practised in carrying the heaviest loads they may be required to carry in action. This is largely a question of knack and training the right muscles.

(d) The Lewis gunner must be taught to cultivate an eye for country, to recognize a good or bad position, to utilize covered lines of advance, and to realize the advantages of a good background.

He requires individual training in the use of ground, and must learn where to place his bipod on uneven ground so that he can fire at a target without further movement, making use of a hillock, for instance, when he requires increased elevation.

He should learn to crawl with a gun for a short distance, resting it upon his leg to keep it out of dust or mud.

He must be taught what is a suitable target for a Lewis gun and what is best engaged with the rifle.

(e) Lewis gunners must learn to move as a team, whether advancing in line to the assault or pushed forward with scouts on a special mission.

(f) Every member of the Lewis gun section must be a good rifle shot and bayonet man, and trained in the use of the bomb.

4. *Control of Fire.*—Everything embraced in the terms fire discipline and fire control is of supreme importance to the Lewis gunner. The Lewis gun is a weapon of opportunity, and premature opening of fire may ruin a well-conceived plan. Ammunition must be husbanded for occasions when it can be used with valuable effect, and fire must cease the moment the target disappears, or when there is no longer need for covering fire.

5. *Indication of Targets.*—Constant practice is required in indication and recognition of targets. Owing to the close pattern formed an error in the point of aim renders the fire of the gun useless. All Lewis gunners must, therefore, be practised in all the common methods of description, whether direct or indirect, by means of reference marks or other aids to indication. They must cultivate a military vocabulary. Fire orders will always be given as orders and in the proper sequence, *viz.*, Range—target (either direct or reference mark—left, two fingers—target)—number of rounds—signal " Fire."

6. *Observation of Fire.*—All numbers must be trained in observation of fire. The small beaten zone of the Lewis gun lessens the difficulty of picking up the strike of the bullets.

7. *Judging Distances.*—Every man in the section must be trained to judge distances accurately up to, at any rate, 700 yards. It is with its first burst of fire that the Lewis gun usually gets its best opportunity

9

of doing damage, and a knowledge of the correct range is essential if this first burst is to take effect. Range-finding instruments are not available for every Lewis gun, but if the gunners learn to use those available during training, and practise as opportunity offers, they will soon learn to judge distances for themselves with accuracy.

8. *Range Cards.*—The preparation of range cards, both for attack and defence, must be practised by every member of the section (*see* accompanying diagrams). A clear, brief, and accurate description must be given of the point from which the range has been taken. No more writing should be put on the card than is absolutely necessary, and everything must be written in block capitals.

Indelible pencil should be avoided because it runs if rain touches it.

If the range has been estimated this fact should be noted.

When a defence range card has been made there may be a tendency to tie the Lewis gun to the centre of the circle, and so restrict its mobility. This should be guarded against; allowances can easily be made for any short move.

9. *Scouts.*—Every unit is responsible for its own protection. All numbers therefore should be trained in the duty of scouts, a pair of whom will precede the rest of the team whenever it moves. The scouts will advance alternately from one piece of cover to another, making good the ground before the gun is brought forward. They will signal the section commander forward to join them whenever they reach a good position for observation, and will receive from him instructions for their further progress, or with regard to the area in which they are to select a fire position. Since the Lewis gun must keep up with the battalion, they must learn to make their reconnaissance rapidly.

10. *Signal Communication.*—Some form of signal communication is of the greatest use to a Lewis gun section. The team will disclose the position of the gun if they bunch round it in action; yet it is impossible to hear orders at any distance when the gun is firing. Moreover, information can be received more rapidly from scouts or observers. Men quickly learn the letters of the semaphore alphabet, sent slowly. They can then, when observing fire, use the signals for correcting range or direction used by machine gunners, and can also invent special codes to meet requirements. Control by the section commander is thus greatly facilitated. Such codes should be uniform throughout a battalion, so that Lewis gun teams can communicate with one another, and should be based on Sec. 164 of Infantry Training.

11. *Tactical Exercises.*—In arranging tactical exercises for a section it is important not to start with anything too complicated; the training should be progressive. But, however simple the scheme, it is imperative that the directing officer should carefully choose the ground beforehand, and make himself thoroughly acquainted with it.

The following makes a useful initial exercise. Place two flags to indicate the position along which the leading line of a platoon has been held up; a third flag will represent the enemy strong point from which the fire is coming. The Lewis gun section, under cover close behind the line of skirmishers, receives orders to push forward to a given position on a flank from which it can cover the advance of the remainder of the platoon.

RANGE CARDS.
Fig. I. Defence.

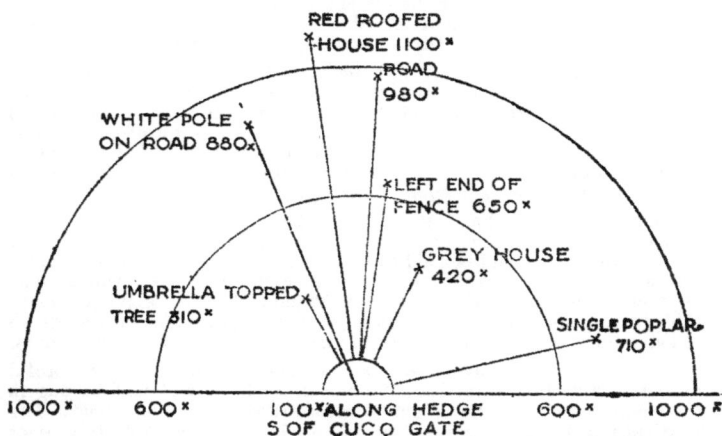

RED ROOFED
✗HOUSE 1100 ✗

ROAD
980 ✗

WHITE POLE ✗
ON ROAD 880✗

✗LEFT END OF
FENCE 650 ✗

GREY HOUSE
✗ 420 ✗

UMBRELLA TOPPED ✗
TREE 310 ✗

SINGLE POPLAR
✗ 710 ✗

1000 ✗   600 ✗   100 ✗ALONG HEDGE   600 ✗   1000 ✗
                          S OF CUCQ GATE

1st April, 1935.   Mean of three ranges,
Pte. John Brown, 42nd Prince's Own.   (Barr and Stroud).

Fig. II. Attack.

An attack range card is made as follows:—

   (i.) Draw 2 parallel lines and fill in *Rendezvous* and *Objective*.

   (ii.) Take or estimate range to *Objective* and write in right-hand column.

   (iii.) Select some object at half-way point and enter its range in right-hand column. Then select and take ranges of intermediate objects. Choose such objects as will easily be recognized when reached and are likely to be near a probably fire position.

   (iv.) A simple subtraction sum will give the range from each successive *Object* to *Objective*. Enter these in the left-hand column and rule out those in the right.

0	First Objective (described).	1700
100	Small Wood.	1600
700	Ruined Farm.	1000
900	Three Trees Centre of Hedge.	800
1300	Mound, Bush on Top.	400
1700	Position of Range Finder (described).	0

Objects described as seen through instrument—should be easily distinguishable when approached.

11

In this practice the following points can be brought out :—

(a) During the advance of the team, movement of scouts by bounds, guiding the section by the best route to the required position; movement of the team in file under the section commander; use of cover; carriage of ammunition.

(b) When the scouts have made good the position, final selection of firing place by section commander; correct mounting of gun without attracting notice, No. 1 crawling if necessary; fire orders and control of team by N.C.O.; disposition of team; scouts out on the exposed flank; observer posted if necessary; all except Nos. 1 and 2 to select positions from which they can at once use their rifles if the gun goes out of action.

Later, three or four sections may be exercised together in consolidating a given line or forming a defensive flank. In such practices special attention should be paid to the co-operation of neighbouring guns. The sections must learn to work together to cover their whole front between them. It is only by constant training that Lewis gunners learn the advantages of enfilade and flanking fire, and gain sufficient confidence to leave the protection of their own immediate front to their neighbours.

12. *Shell-hole Tactics.*—It is also possible to carry out useful exercises in shell-hole tactics. Groups of shell-holes should be dug in the training area, and Lewis gun sections ordered to occupy them. The men must learn to recognise at a glance which hole is likely to present the most valuable field of fire for the Lewis gun, while other numbers occupy other shell-holes to give it protection on a blind side. The conversion of a shell-hole into a gun position should be practised (*see* S.S. 202, Organization of Shell-hole Defences). When one group of shell-holes has been occupied a rapid move from one group to another may be practised, scouts creeping out to reconnoitre before the gun is brought forward.

13. *Live Ammunition.*—Suitable attack exercises should occasionally be carried out with live ammunition. Targets can be constructed roughly to represent dummy figures or the loophole of a pill-box. The men gain confidence by seeing the results of their fire.

14. *Gas Alarm.*—At certain stages the gas alarm should be sounded, to practise the gunners in aiming through gas masks.

15. *Discipline.*—At all times, both in and out of the line, rigid discipline must be enforced. Unless this is done, it will not be possible to control the team in actual fighting. Each man in the section must be made to realise that his work is just as important to success as that of the man actually firing the gun.

IV.—LEWIS GUNS IN THE OFFENSIVE.

A.—The Trench to Trench Attack.

1. *Before the Attack.*—Before an attack the task of Lewis guns consists in preventing the enemy from repairing damage to his wire or front defences, and in supporting any raids which may be ordered. It is inadvisable to send the Lewis guns actually to accompany the raiding party into the enemy's line, for they are liable to delay its movements, are in the way in the case of close fighting in the trench, and, at night, cannot distinguish friend from foe. They are better used to cover the raiding party while it crosses and re-crosses No Man's Land. For this purpose Lewis guns should be secretly pushed out beforehand into positions on either side of the route to be taken by the raiders. Certain guns will have special instructions to watch any suspected machine gun emplacements, while others will be sited to sweep between them, the whole length of the enemy's line. They will remain in position till the raiding party has returned.

2. *During the Attack.*—During the advance to the first objective, covering fire may not be required from Lewis guns. If, however, this objective is at a considerable distance it may be useful to push Lewis guns forward to silence any machine guns or snipers. Such guns, whether drawn from the first or later waves, can rejoin their platoons as they pass.

Another method of covering the attack is to fire Lewis Guns from a sling while advancing. (*See* Appendix I., "Firing from the Hip.")

During the advance Lewis guns will be employed to cover gaps occurring laterally or in depth, and to deny them to the enemy until closed by the advance of later waves. If one company gains ground while the next is held up, the Lewis guns of the former can enfilade the trenches which still hold out, and can assist bombers working down them by covering them from attack over the top.

If a platoon is held up by a machine gun the Lewis gun will engage it at once in order to put it out of action or unsteady its fire. In any case it will hold the machine gunners' attention while other sections work round the flank.

3. *After the Attack.*—It is immediately after a successful attack that Lewis guns find their greatest opportunity. Commanders therefore should not commit too many guns unduly in the earlier stages of an attack, and should retain in hand a number sufficient to carry out the important duties that devolve upon them.

These duties may be divided under three main heads, and Lewis gun sections should be definitely allotted beforehand for each class of work; (*a*) organizing the ground won; (*b*) covering this organization; (*c*) exploiting the success.

13

(a) On reaching the general line to be organized, a proportion of Lewis guns can usefully be employed in echelon to the rear to form a defensive flank, until this is secured by the advance of neighbouring units.

When the organization of the captured objective begins, short lengths of trench will first be dug, and in these Lewis guns will be so placed that they can mutually support one another, and put a complete belt of fire across the front of the position. Other Lewis Guns will be allotted to the garrisons of strong points in rear of the line, so that an enemy counter-attacking in force may be opposed at every step; these may be replaced by Vickers guns when the latter come up to complete the defensive organization.

(b) Lewis guns will be of value to cover the work of organizing the defence by occupying posts pushed out in front of the line. The two or three men with each gun can creep forward unseen and find cover in shell-holes, from which they may find numerous opportunities for fire at the disorganized enemy, and from which they can open fire on an enemy moving forward to counter-attack. The guns allotted to this task should not, however, be pushed forward into isolated positions, and should be provided with an adequate number of riflemen as a protection against surprise or being rushed when the gun is out of action owing to stoppages.

The principle that a unit is responsible for its own protection from all directions includes protection from the air. A certain proportion of Lewis guns should be allotted (either under battalion or brigade arrangements) to deal with low-flying enemy aeroplanes which attempt to cross the captured area; these should be sited in the best available cover, not more than 800 yards behind the line. (See Section VII.)

(c) The rifle, with its bullet and bayonet, is pre-eminently the weapon for killing any enemy slipping away over broken ground, while the bombers clear underground shelters; but the Lewis gun will also have its opportunities. In a difficult situation the enemy is often inclined to bunch, and this may afford the Lewis gunners a favourable target.

As soon as the captured area is cleared, parties will at once be sent forward to ascertain the enemy's new dispositions and seize any tactical points whose possession is of importance. If these parties are supported by Lewis guns their advance can be more bold and rapid, and their hold on any ground gained more secure.

(d) Another task which in the later stages of a battle may fall to the Lewis gunners on reaching a final objective is to prevent the removal of enemy artillery. Enemy guns may have been located within range, in which case a special party, including one or more Lewis guns, should be detailed to keep the guns under observation and prevent their removal until they can be destroyed or captured.

B.—In Open Warfare.

4. The normal method of advance in open or semi-open warfare will be " by bounds," each unit sending forward a proportion of its force to secure a definite point before committing the remainder. The

formations of the leading infantry will vary according to the ground and the tactical situation, but as a rule it is better to keep the Lewis gun in close support of, and not actually in, the first line. If the Lewis gun is in rear, the platoon commander has greater freedom in its use, and can either push it straight into his firing line, if immediate covering fire is necessary, or work it to a flank to get enfilade effect and assist an enveloping movement.

5. The opposition to be expected is of various kinds. If the enemy has left small mobile parties to cut off our advance parties and hinder reconnaissance, our own advance parties must be sufficiently strong to defend themselves till reinforcements arrive, and must arrange for neighbouring parties to give mutual support. If such a party consists of at least two sections, one armed with a Lewis gun, it should be strong enough for this purpose; and it will have sufficient fire power to cover the advance or withdrawal of a neighbouring patrol.

More frequently, definite positions carefully chosen and organized for defence will be encountered; these may range from small strong points held by a machine gun or a few automatic rifles to strongly defended localities, such as woods and villages, or a certain length of line sited in a favourable tactical position.

The use of Lewis guns in attacking each class of position will vary with the conditions, but three considerations are of universal application:—

(a) There must be no hesitation in the execution of a bound. Resistance must be removed quickly, in order not to delay the bound of units behind.

(b) The enemy positions will have definite flanks, unlike a trench-to-trench attack. Full advantage must be taken of this to employ enveloping tactics.

(c) The enemy should not be allowed to escape, and arrangements for cutting off his retreat should be included in the orders for the attack.

6. The reduction of a small strong point, which may be garrisoned by a few riflemen or by a single machine gun, gives a platoon commander an opportunity for exercising his tactical skill. These points will occur at the most unexpected places, so that the commander on the spot must make his dispositions on the spur of the moment, without waiting for instructions. By slowness in appreciating the situation, or by reckless handling of his sections, he may imperil his whole platoon, whereas prompt and intelligent action would have captured the position with little or no loss.

In attacking such a position the Lewis gun can be employed in two alternative ways. If the platoon is caught by machine gun fire in an unfavourable position, the Lewis gun will engage it at once to cover the movements of the other sections which will endeavour to get forward to the flanks. If, however, the strong point is discovered before the platoon comes under heavy fire, it may be advisable to send the Lewis gun to a flank to some position from which it can continue to give covering fire up to the latest possible moment, and can also annihilate the retreating enemy if he is forced from his cover by bombs or rifle-bombs or driven out by the bayonet. On the reduction of the strong point the Lewis gun will at once move up to help to defend the ground gained.

7. In the case of larger centres of resistance, more difficulty may be experienced in reaching assaulting distance. In such cases covering fire from Lewis guns will assist the remainder of the force to get forward more safely. If, for instance, some bare patch of ground has to be crossed within range of the enemy's position, Lewis guns can often be pushed forward unobserved in front to neutralize the enemy's fire at the moment the attacking line becomes exposed.

Lewis guns, again, change front in a moment, and can be temporarily detached to deal with any body of the enemy which attempts to check the advance by any action from a flank.

The Lewis gun plays a valuable part in the fight for superiority of fire before the assault, and some should be sited in positions from which they can continue to give covering fire up to the last possible moment.

8. *Woods* organized for defence will frequently be encountered as centres of resistance, and enveloping tactics will often be found best suited for their reduction. It is advisable to push past the wood on one or both flanks before actually attacking it; if covering fire from field or machine guns be laid upon the sides of the wood this can usually be done with little loss. Lewis gun posts will then be sited on the side of the wood and can give very valuable support to the troops that eventually make a frontal attack; they can also cut off the enemy's retreat and prevent the arrival of his reinforcements.

In the case of large areas of woodland which cannot all be captured in one bound, Lewis guns may usefully be sited to sweep all clearings. By this means they can isolate the sector first selected for attack, and prevent the enemy from moving troops rapidly from some other sector to the threatened point.

The troops actually advancing through a wood must use their Lewis guns according to circumstances, but they must always be used boldly, and a good proportion of them kept well forward.

As a general rule it is advisable, when entering a thick wood, to keep the Lewis gun team in file close behind the first line of skirmishers. If the Lewis gunners actually form the skirmishing line there is likely to be delay in replacing a casualty on the gun or supplying it with ammunition.

If the front line is held up by an invisible enemy firing through undergrowth, the Lewis gunners can pass through and advance firing from the hip. Similarly, if a line of wire is met with, the Lewis gunners will pass through the first gap and keep up covering fire while the obstacle is removed.

When an entire wood has been captured it is wise to push forward from it as soon as possible. The line should be consolidated in front of it, and no time should be lost in getting Lewis guns forward to their positions.

9. The same enveloping tactics should also be employed in dealing with defended *villages*, Lewis guns being sited on the flanks wherever they can fire down cross streets and command the exits.

To cover an actual house-to-house attack a couple of Lewis guns, fired from behind corners, can make a narrow street impassable. Mounted behind heaps of rubble, or occasionally in windows, they can supply supporting fire wherever it is required, moving forward rapidly as each successive house is taken, and choosing new positions to cover the next movement or cut off the enemy's retreat.

10. In any form of attack the supply of ammunition to Lewis guns requires the most careful consideration. The number of the magazines actually carried by the section will always depend upon the circumstances. In open warfare, where the advance may be of several miles, 21 magazines per gun will probably be as much as a leading section can carry. In an attack from trenches with a limited objective 30 magazines per gun may be required, because this number can be carried by the section for a short distance and the Lewis guns may have to fire quickly to stop an immediate counter-attack. In any case a certain proportion of the magazines should be kept in reserve either on the limber, in dumps, or with a carrying party, in order that damaged magazines may be quickly replaced.

V.—LEWIS GUNS IN THE DEFENCE.

1. In all defence problems the first consideration of a commander is, "How many men do I need to make this position reasonably secure?" Whatever the type of defence, no more men than are actually necessary to hold the position should be employed.

To render a line reasonably secure it is necessary to be able to bring heavy fire to bear upon all ground immediately in front of it across which an enemy can attack.

These two principles emphasize the value of automatic weapons in defence.

A.—In Trench Warfare.

2. In the defence of a highly organized trench system, Vickers or Lewis guns will be sited so as to provide belts of cross and converging fire across the front. The scheme of the Lewis gun defence must be made to conform to that of the Vickers guns, so that there may be no overlapping and waste of fire-power. (S.S. 192, "The Employment of Machine Guns," Part I., Sec. 17, para. 1.)

If the ground were level and the Vickers guns could be safely sited in ideal positions, all the front could be swept by their sustained fire: but it is seldom that such conditions are found, and small gaps will generally occur. A brook or sunken road may cross the line, affording a covered approach to the enemy, over which the Vickers gun line of fire will pass. There may be a depression in front of the line within bombing distance—dead ground to the low-sited Vickers guns. Again, there are often positions with good fields of fire where an elaborate machine gun emplacement would be conspicuous. In all such positions Lewis guns will be needed to supplement the Vickers guns, and their arrangement can only be decided by the Infantry officer commanding the sector in consultation with the officer commanding the Vickers guns. (*See* accompanying figures.)

While the main line of defence will thus be based upon the positions chosen for the machine guns, there will normally be in front of this main line a forward line, consisting of a line of small posts or short lengths of trench, in order to prevent the enemy from observing and reconnoitring the main line, to keep him under close observation, and to break up an attack and give warning to the troops behind. In this line a certain number of Lewis guns are of the greatest value. These Lewis gun posts can be sited in places where they are least vulnerable to enemy shell-fire and at the same time can enfilade an attacking enemy, disorganize his line, and cause him to fall behind his own creeping barrage.

3. All Lewis guns detailed for the defence of any position must have assigned to them definite areas which they have to watch, such areas being usually diagonal to their front. Unless this is done, in case

of a general attack, all guns would probably open fire on the most prominent body of the enemy, thus permitting other parties to get forward unperceived.

Only in special circumstances will such guns be allowed to fire in other directions. If "A" sector be raided and "B" sector be not threatened, the guns defending "B" may temporarily swing round to assist in defending "A." But in all such cases one man of each team must be detailed to watch the original line, so that the gun may be swung back the moment a target presents itself in that area.

Alternative positions must be selected for each gun, from which it can sweep the front which it has to defend, for the Lewis gun finds safety rather by rapid change of position than by the strength of its emplacemnt.

V = Vickers Gun.
L = Lewis Gun.

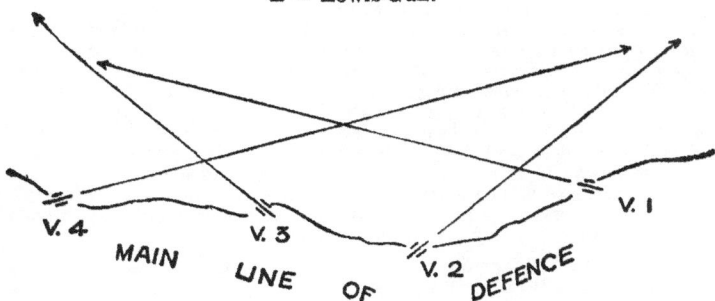

Machine guns putting up belts of fire in level country.

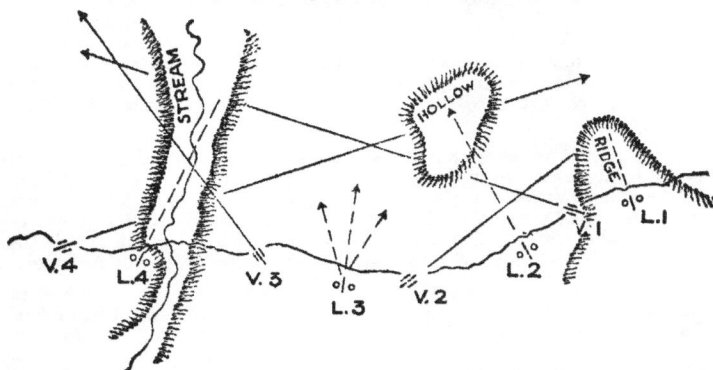

Lewis guns assisting Vickers guns in main line of defence, when the Vickers gun fire is interrupted by ridges, hollows or watercourses which leave dead ground, or where there is a long gap between guns which the enemy might cross at night. (L. 3.)

19

4. Owing to this mobility of the Lewis gun and the absence of a fixed platform, " emplacements " in the ordinary acceptation of the term are not necessary, but definite firing-places will always be constructed. In most cases all that is required is a depression running diagonally across the parapet; a diagonal firing-place is less conspicuous, as the gunner gets a certain amount of head cover, and the gun is firing in the best position to defend the line. All firing-places should be protected from enfilading shrapnel by a strong traverse built on the side from which the prevalent shell-fire comes.

Every gun team should, where possible, have a strong dug-out adjacent to its firing-places; but, in addition, a shelter for the gun must be made at every firing-place by sinking a box into the side of the parapet near the bottom of the trench.

Similar recesses should be made for the magazines, which should be distributed at different firing-places; not more than eight magazines at most should be kept in one recess, so that no more than that number can be buried by one shell. The number of magazines per gun to be taken into the trenches depends upon circumstances, and will be decided by the battalion commander, after consultation with his Lewis gun officer. Thirty per gun is a usual number, some being kept in reserve at company or battalion headquarters. A proportion of the magazines should be emptied and cleaned thoroughly every day, and any damaged ones replaced; if this is done no difficulty will arise in regarding the bulk of them as trench stores, to be handed over on relief, and much labour will be saved to relieving gunners. A box of 1,000 loose rounds should be kept in every Lewis gun dug-out.

Both gun and ammunition recesses should be protected by a gas curtain. Gas affects cartridges much more rapidly than it affects the gun, covering them with a green slime that at once causes stoppages; magazines that have been exposed to gas should, therefore, be emptied at once and refilled with clean rounds. If the ammunition be clean no immediate difficulty will be experienced in firing. The gun should be cleaned and oiled at the earliest opportunity, and all parts well washed in boiling water containing a solution of washing soda. Till this has been done corrosion will continue, and in a few hours will cause stoppages.

5. A belt of wire should be maintained in front of each line of posts, to prevent enemy scouts from creeping between the posts, to keep his bombing parties out of range, and to delay assaulting troops under the close fire of the automatic weapons. The wire should, therefore, be not less than 50 yards from the line, and where possible should be placed on the near side of a bare stretch that affords little or no cover.

By the exercise of a little ingenuity the wire can be placed in such a way as to cause the enemy to bunch in the required line of fire. This is effected if the wire is bent back between the posts so that the Lewis guns or machine guns can enfilade the enemy's side of it. Care must, however, be taken that it is not laid too obviously. If some of the entanglement consists of low wire in long grass, or in shell-holes, and minor irregularities are introduced, this danger is avoided.

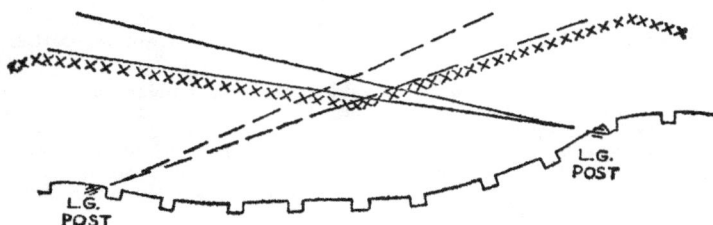

Such an arrangement has great value at night, since the moment an enemy is heard touching the wire, the Lewis gun can fire with every expectation of hitting him, at whatever point he may be. Any gap in the entanglement, or a narrowing re-entrant in it, will tend to make an attacking enemy bunch. But if a gap is left thus as a trap it must be filled up on relief, or its position must be carefully pointed out to relieving gunners.

6. Though occasionally two or more Lewis guns may have to be grouped close together in accordance with the general scheme of defence, as a rule it will be possible to keep Lewis gun sections in their own platoon areas.

Every man of a Lewis gun section holding the line takes his turn of duty on the gun; he also does his share of cleaning guns, magazines, and ammunition, and construction and repair of firing-places and shelters; so that this section cannot be expected to undertake more work or find additional posts. At the same time, the gun will sometimes need additional protection on a blind side. At night, again, in certain circumstances, it may be necessary to strengthen the sentry posts from other sections for the protection of the gun. Night patrols and wiring parties should also be found from other sections. In the confined space of No Man's Land there is little utility in sending out Lewis guns actually to accompany patrols. But while patrols are out it is often useful to push reserve Lewis guns to lie out in front of the line to support them in case they are attacked, and, in particular, to prevent their being cut off by hostile patrols during the return journey.

7. As soon as a battalion takes over a new line, all Lewis gunners will be practised in the duties they will have to perform in case of any emergency. Every gunner must know the ground his gun has to cover in case of an attack, and all must be constantly exercised in bringing their guns quickly out of their dug-outs into action at their firing-places. Practice is also needed in filling magazines rapidly in a dug-out.

Sections in reserve must learn their way about all the communication trenches, and know the quickest way between any two points. All sections behind the front line should construct some form of anti-aircraft mounting for use against low-flying aeroplanes.

21

Lewis gunners must take every opportunity of reconnoitring the ground in front of the trenches, to find any convenient hollows or shell-holes from which they can protect our own patrols or working parties, and harass those of the enemy.

Constant attention must be paid to the proper observance of the points before, during and after firing. After firing, it may not always be possible to strip the gun for some hours; but this must be done at the earliest opportunity. Care must be taken that neighbouring guns are not stripped at the same time; if the situation demands such a precaution, a reserve gun may be brought up to take the place of one that is being cleaned.

B.—In Open Warfare.

8. Defence in open warfare is based on exactly the same principles as in trench warfare. A few Lewis guns in an outpost line will stiffen its power of resistance and economise men. There is a greater choice of ground, and an eye for country becomes of the highest value. In trench warfare there is little room for doubt as to the direction and method of attack. In open warfare, not only may the attack be other than frontal, but the enemy cannot start from ready-made cover within assaulting distance. The commander of a defending force must, therefore, consider the enemy's probable point of deployment, the direction of his advance, and the places where he may try to build up his line for the assault.

His appreciation of the situation should be based on reconnaissance of the ground, aeroplane reports and photographs, and the study of maps.

Automatic weapons should always be sited to sweep any ground where the enemy is likely to form up for the assault. This will often be dead ground from the main line of resistance, so that Lewis guns, pushed forward well in front, may be necessary. Such guns should lie concealed and reserve their fire till the moment when they can inflict the most damage; it is often possible to site them to enfilade some bank or ditch that seems likely to invite the enemy to its shelter. If the enemy fails to act as expected, the mobility of the Lewis gun enables it to shift its position to meet the new situation, or to regain the main line.

Rifle fire should be used to stop the screen of enemy scouts and prevent the premature disclosure of the gun.

9. Lewis guns should support the patrols sent forward to watch positions where the enemy is considered likely to deploy, and in favourable circumstances they may often have a chance of surprising formed bodies of the enemy before they expect opposition. This can most frequently be effected either when an attack is expected at night, or when an attacking force has to pass through broken or wooded country. In either case an attacking force will delay its deployment as long as possible in order to avoid losing control or direction.

A Lewis gun used successfully in such a way has not only material, but great moral effect. The attacking force is badly shaken at the outset. If a company, debouching from a wood, is caught in close formation by Lewis gun fire, the men will probably rush back to cover in disorder, and disorganize the troops behind; at night the attack will probably never materialise. It is therefore worth taking considerable risks to gain such an effect.

Lewis guns should particularly watch any natural features that are likely to cause an enemy to bunch, such as bridges, fords, defiles, causeways between pools of water, tracks through woods, or even the easiest place to get over a hedge, and they should be ready to enfilade any obstacle which is likely to deflect the enemy's line of advance.

By such means the Lewis gun is likely to obtain the deep dense target which allows it to develop its maximum effect.

10.   In the main line of defence, positions which give a wide, long field of fire should be distrusted. Such a position is suitable as an O.P., but not as a site for a gun. A well-concealed gun with a shorter field of fire is likely to survive longer and do better service. Apart from any question of casualties, a Lewis gun which begins to fire at an advancing enemy at extreme range will be too hot to fire when the enemy is within really effective range

VI.—CO-OPERATION.

1. The various weapons of an army are complementary to one another, and success can only be obtained when all work together intelligently for a common object.

To co-operate successfully it is necessary that a man armed with one weapon should have a working knowledge of the other weapons with which he comes into contact, so that he may realise their capabilities and limitations.

2. All battalion weapons—the rifle and bayonet, Lewis gun, bomb, and rifle-bomb—are represented in the platoon, and the men handling them are, therefore, taught to work together from the earliest stage of training.

The rifle combined with the bayonet is the main weapon of every infantry soldier, and is essentially the weapon for offensive action. The bomb and rifle-bomb supplement the action of bullet and bayonet, and are designed to kill the enemy below ground or force him into the open, where he may offer a target to the rifle and the Lewis gun. The Lewis gun assists the rifleman to get forward by applying heavy covering fire instantaneously in any required direction, and is supreme among battalion weapons in its power of stopping an attack by a mass of men across a narrow space.

When a short length of trench has to be taken by a bombing attack, the Lewis gun. can assist by keeping down enemy rifle or machine gun fire by preventing the bombers from being rushed over the top, or by stopping further supply of bombs to the enemy; and when a trench block has been made a Lewis gun may usefully be posted on a flank to guard it. Bombers, in their turn, can guard the Lewis gun from attacks by enemy bombers creeping up saps or through shell-holes towards its position.

The rifle again co-operates with the Lewis gun in defence by engaging any targets unsuitable to the gun, such as a scattered line or single scouts and riflemen.

3. Lewis guns must not only co-operate with the other battalion weapons, but also with one another. Company commanders should bring out the importance of this in company training, and prevent platoons becoming too self-centred.

If No. 1 and No. 3 platoons are attacking side by side, instances will frequently occur when the Lewis gun of No. 3 platoon can support the advance of No. 1 platoon much more effectively than can the Lewis gun of No. 1, and *vice versâ*. Similarly, in defending a line a Lewis gun will often be sited to protect the front of a neighbouring platoon, and will need protection for its own front in turn. Again, when two or more guns are grouped to undertake a special task, a complete understanding is necessary on all questions (such as points of aim, occasion for opening fire, etc.) if full effect is to be obtained from all guns employed. Definite instructions must be given before the sections set out, and complete arrangements made for inter-communication.

4. There are two other weapons which work closely with infantry battalions, but belong to different organizations—the Vickers gun and the light mortar. The company commander should maintain close touch with the officers commanding these weapons in his sector. He must remember that they cannot assist him unless he keeps them fully informed of the doings of his company, and discusses with them beforehand any contemplated enterprise. The siting of the Lewis guns depends to a great extent on the siting of the Vickers guns, and cases will frequently occur where Lewis guns and light mortars, working in conjunction, can inflict much heavier losses on the enemy than either could effect by itself.

The co-operation of Lewis guns with Vickers guns in defence has already been considered. (Sec. V., paras. 2 and 3.) In attack the Lewis gun is in a sense the forerunner of the Vickers gun. In the preliminary organization immediately subsequent to a successful attack, Lewis guns perform the function of Vickers guns, so far as their nature permits; but when the Vickers gunners follow up, the Lewis gunners must never hesitate to surrender to them any positions they select. The Vickers gun can make better use of a good position, and the Lewis gun has a wider choice of positions open to it.

5. To escape bombardment, the enemy frequently leave their trenches and form posts in shell-holes in front of and behind the line. The light mortar, provided an adequate supply of ammunition can be ensured, is the best weapon with which to dislodge them. But before it opens fire all the Lewis guns in the sector should be advised of its programme, so that they can cover the shell-holes and prevent the escape of their occupants. The mortar can then range deliberately till it secures direct hits on each target. Similar tactics may be employed against a strong point encountered during an attack.

VII.—THE EMPLOYMENT OF LEWIS GUNS AGAINST AIRCRAFT.

1. Every Commander is responsible for the protection of his own command against surprise. A force can only be regarded as secure from surprise when protection is furnished in every direction from which attack is possible.

The enemy does, and will, attack from the air; therefore every unit must provide its own protection against aerial attack. Moreover, enemy aircraft flying below about 3,000 feet can only be dealt with by small arm fire, as our own aeroplanes and anti-aircraft guns cannot attack successfully enemy aeroplanes flying at these low altitudes. Lewis guns are one of the chief weapons to be used against these aeroplanes, and Lewis gunners must be trained accordingly. To carry out this work adequately they must be provided with anti-aircraft back and foresights, and instructed in their use.

2. The enemy will use low-flying aeroplanes for :—

(a) Contact patrol work.

(b) Harassing troops in forward area with machine gun fire and bombs.

(c) Bombing billets, camps, dumps, railheads, etc.

(a) *Contact patrols* are employed by the enemy both in defence and offence. They act as a means of communication between the front line troops and headquarters. Their work is not so much fighting as receiving and transmitting messages from the infantry, watching their progress and directing gun fire on important points.

During a battle it is essential that the enemy's contact patrols should be harassed as much as possible, and fire should be opened by all guns within range, so as to keep the hostile machines at a distance.

(b) *Harassing fire.*—To deal with this, the forward area should be covered by several lines of guns. At 3,000 feet one gun can cover effectively a circle with 500 yards radius; therefore, one gun per 500 yards of front is required to provide a continuous belt of fire.

Greater concentration of fire is obtained on any one spot if guns are grouped in pairs at not greater intervals than 1,000 yards. Therefore, to cover effectively the forward area as far back as the line of the field guns, three lines of Lewis guns will be required :—One about 500 yards behind the front line, a second about 1,000 yards in the rear of the first, and a third about 1,000 yards in rear of the second, or 2,500 yards behind the front line.

If more guns are available, a fourth line should be established about the line of the Heavy Artillery.

This method will ensure that no enemy aircraft can fly over our forward area at a height less than 3,000 feet without being engaged continuously by at least two Lewis guns

This system will deal with all aircraft coming within categories (a) and (b).

All Lewis guns employed in this system must be equipped with anti-aircraft sights and mountings, but they must also be prepared to engage ground targets in an emergency.

(c) *Bombing billets, camps, dumps, etc.*—Enemy aircraft which have crossed the forward area at a high altitude and have succeeded in avoiding our own aeroplanes and anti-aircraft guns, may then dive to a low altitude for the purpose of bombing or harassing with machine gun fire.

Protection must therefore be provided for billets, camps, dumps, etc., by means of Lewis and machine guns. These guns, together with the anti-aircraft sights and mountings, will be provided by the units occupying the billets or camps: in the case of important dumps, railheads, etc., special units will be detailed to find the guns and accessories.

In arranging the scheme of defence, the following points must be taken into consideration:—

(i.) An aeroplane travelling at 80 miles per hour and at an altitude of 3,000 feet has to release its bombs 500 yards before reaching a point vertically over its objective: consequently the guns should be placed about 400 to 500 yards outside the locality to be defended.

(ii.) The line of fire of each gun should be such as to ensure that as far as possible spent bullets fall in open country and not on other billets, towns, roads, etc., frequented by troops or civilians.

The scheme of defence for each locality should be kept by the Town Major, Camp Commandant or other responsible officer. Each unit on coming into the locality should receive orders from this officer as to the number of guns to be found by the unit and the positions to be occupied.

(d) Corps and Divisions must prepare and co-ordinate the schemes of defence for (a), (b) and (c): a Staff Officer should be responsible for the co-ordination of the defence in each Corps and Divisional area.

3. The *training* of Lewis gunners for anti-aircraft work includes such a variety of subjects that it will probably be found impossible to have more than a comparatively small number with all the necessary information at their command. In the choice of those who are to receive more than the merest mechanical training, the importance of strong eyesight and good hearing should not be forgotten.

The *elementary* instruction which should be given to every Lewis gunner includes the use of anti-aircraft sights and the principles on which their design is based, also the types of mounting they are likely to meet, from permanent mountings in back areas to portable ones in attack. The importance of continuous fire in anti-aircraft work, as distinguished from the " short-burst " method employed against ground targets, should be carefully taught. Aiming on model aeroplanes should familiarize the pupil with the use of the sights, while drill, which includes getting into action against aircraft, and from aircraft to infantry, should render him capable of quick and confident handling of the gun, sights and mounting. An exhibition flight at given heights and speeds arranged at an aerodrome would be an invaluable help.

More *advanced* training should include lectures on the following subjects:—

(i.) The employment of Lewis guns against aircraft.

(ii.) A pilot's experiences of small arm fire.

(iii.) Types and speeds of aeroplanes. Every opportunity should be given to train gunners in recognition of the various types. This can best be done by visits to aerodromes.

(iv.) Aeroplane tactics.

APPENDIX I.

FIRING THE LEWIS GUN FROM THE HIP.

1. The possibility of firing the Lewis gun from the hip, in the same way as the French use their Chauchart automatic rifle, was demonstrated in the early days of the Somme offensive; and during 1917 a number of instances have occurred of its successful and unsuccessful use, both with and without a sling. The Lewis gun, fired from the hip, is less effective than the Chauchart—a weapon which is about half its weight, fires more slowly, and has more portable ammunition—but at times it has undoubtedly proved useful.

2. The advantage of being able to produce heavy fire while moving is obvious, both in covering an attack and in pursuit. But this use of the Lewis gun is open to several disadvantages, and must be used with great discretion.

These disadvantages are as follows :—

(i.) The aim is *much* less accurate. A Lewis gunner, firing from the ground, ought to be certain of getting effect at once at close range, and at any range up to 300 yards he should put a proportion of his shots into the loophole of a pill-box before he has emptied his first magazine. When firing from the hip with a sling, a highly-trained man would be fortunate to get a single bullet into a loophole during an advance over broken ground. Though material effect has occasionally been reported, the value of this form of firing must be regarded mainly as moral.

Tests on a range may be misleading, and, though a number of men can make good practice while advancing at a target to their front which is visible all the way, it is infinitely more difficult to bring fire quickly to bear on an enemy appearing in an unexpected place amid the distracting circumstances of actual fighting.

Since fire can only be applied with any approach to accuracy in the direction in which the gunner is advancing, this method offers no protection against the machine gun firing from a flank, which is usually defiladed from its own front.

(ii.) It makes the gunner very conspicuous. Where strong opposition is met and casualties are severe, the Lewis gunners may suffer heavily. Thus, in the event of an initial failure, there may be no guns placed to cover the withdrawal, or hold on to such ground as has been won.

(iii.) In theory, a Lewis gunner should be able to fire his gun throughout the assault, shoot the enemy in the trench when he reaches it, and then take part in the consolidation. But changing magazines or remedying stoppages from this position makes it difficult for the gunner to keep abreast of the leading line, while if he does fire continuously he may be encroaching dangerously on his supply of full magazines, and his gun may be getting hot before he has to meet a counter-attack.

3. To sum up—The Lewis gun can be fired from the hip without a sling for a short distance; the employment of the sling undoubtedly lightens the task of the firer, but the sling must be capable of rapid attachment or detachment.

Material effect is uncertain, but the moral effect is considerable, and, where sufficient covering fire cannot be applied in other ways, this method may prove of value in keeping down enemies' heads. It is likely to be most useful when an enemy is retiring or when he has only cover from view, as, for instance, when firing at our troops from undergrowth in woods or from standing crops. It must therefore only be used where special circumstances demand it, and no one but the commander on the spot is able to decide on the right time for its introduction.

APPENDIX II.

AMMUNITION.

1. The establishment of S.A.A. earmarked for Lewis guns is as follows:—

With the Battalion .... 4,318 rounds per gun.
In Divisional Ammunition Column ... 2,000 ,, ,, ,,

2. The battalion supply consists of 33,088 rounds, loaded in magazines (44 magazines per gun) and 36,000 rounds of loose S.A.A. in clips in 1,000-round boxes. This is carried in four limbered G.S. wagons, of which one carries the guns and ammunition for each company.

3. The load of a Lewis gun limbered G.S. wagon is as follows:—

	lbs.
Four Lewis guns, @ 27 lbs.	108
Four sets spare parts, @ 15 lbs. ......,..............	60
176 magazines filled, @ 4½ lbs.	792
Four sets spare parts, @ 15 lbs.	60
22 tin boxes, @ 8¼ lbs. (each holds 8 magazines in 2 canvas carriers or 4 webbing pouches)...	182
9,000 rounds S.A.A.	675

1,949 lbs.

4. Pioneer battalions, having 8 guns, have only two Lewis gun limbered G.S. wagons, with a proportionate supply of ammunition.

5. No definite orders have been issued with regard to the packing of the limbered wagon, but the following method has been found serviceable:—

In the front half:—2 guns with spare parts, 11 tin cases containing 8 magazines, 4 boxes S.A.A. (1,000 rounds each).

In the rear half :—2 guns with spare parts, 11 magazine cases, and 5 boxes of S.A.A.

The guns are in their wooden chests, raised on wooden projections, which form a shelf down each side of the wagon. This shelf is of such a height that the fastenings of the boxes are clear of the top of the side, to facilitate unpacking.

On the floor are distributed the S.A.A. boxes, 2 down either side under the gun boxes, and in the rear portion, 1 across the front. Between are placed 3 magazine cases, and 2 spare parts bags lie 1 at each side next the tailboard.

A second layer is formed by two rows of 3 magazine cases between the guns, standing on their longest side with handle uppermost. Between these and the tailboard 2 more magazine cases are up-ended on the floor.

A plan of the rear portion is attached.

Since the front portion has 1 less box of S.A.A. than the rear, it can carry the driver's pack, spare harness, etc., and also a receptacle for any spare parts, tools, instruments, etc., which are not required to be taken with the gun in any particular operation.

PLANS OF REAR PORTION OF G.S. LIMBERED WAGON.

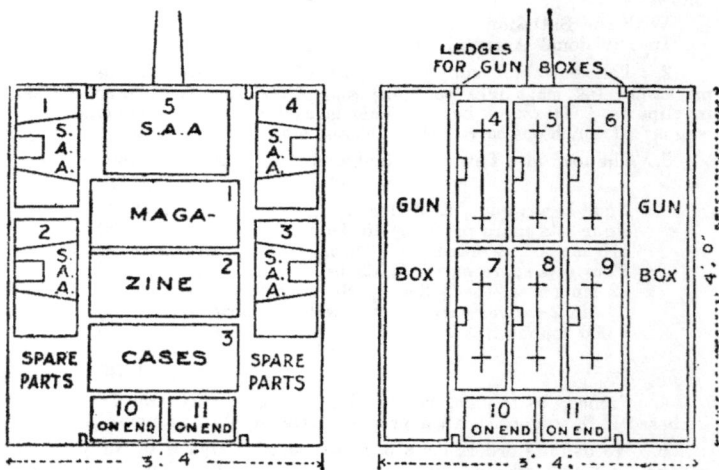

LOWER LAYER.   UPPER LAYER.

Total amount of ammunition carried in rear portion :—

88 full magazines in canvas carriers, packed in 11 tin cases; 5,000 rounds in clips in 5 S.A.A. boxes.

Front portion contains the same number of guns, spare parts bags and magazine cases, but only 4 boxes S.A.A. instead of 5, thus leaving room for drivers' pack, spare harness, etc.

APPENDIX III.

LEWIS GUN DRILL.

1. Lewis gun drill can never take the place of ordinary close order drill in the teaching of discipline, nor as physical exercise. Lewis gunners must, therefore, frequently take part in drill (with rifles) with the rest of the platoon.

Lewis gun drill is, however, an essential part of elementary instruction in the gun. By its means the beginner is taught to adopt instinctively the correct firing position, to mount, load and lay the gun quickly and correctly, and to understand fire orders, whether spoken or signalled. Each man takes his turn in handling the gun.

When the squad is proficient in drill on the parade ground, the lessons learnt should be applied in range practices, and in small tactical exercises in which the gun is brought into action on rough ground.

2. *Signals.*—The following signals are used by the Lewis gun commander in controlling fire:—

Hand up = Prepare to open fire.
Hand dropped = Open fire.
Elbow close to side, forearm waved horizontally = Cease fire.
Arm swung in circular motion in front of body = Out of action.

3. *Position of Gun and Stores.*—For elementary drill, the Lewis gun is placed on the ground resting on butt and bipod, with canvas cover fastened over working parts. The spare parts bag and a magazine carrier containing four magazines are placed on the left of the gun at two paces interval.

4. *To Fall In.*—On the command " Fall in," the squad falls in in single rank, five paces in front of the gun.

On the command " Number," the squad numbers off from the right.

5. *To Take Post.*—On the command " Take post," the squad turns to the right and doubles round behind the gun.

No. 1 (*a*) Takes up position on left of gun;
   (*b*) Removes canvas cover and examines gun;
   (*c*) Takes magazine from No. 2 and places it on gun;
   (*d*) Reports " Ready."

No. 2 (*a*) Takes up position on left of carrier;
   (*b*) Puts on spare parts bag and examines magazines;
   (*c*) Hands one magazine to No. 1 and closes carrier.

After this, No. 1 repeats all words of command.

6. *Action, Range, Target.*—On the command " Action, Range, Target " (the instructor giving details of the range and target)—

No. 1 (*a*) Adjusts sight to range ordered and then lowers leaf;

(*b*) Runs forward with gun, right hand grasping the small of the butt and left hand under the radiator casing; on reaching the position indicated by the gun commander, he throws the gun forward with both hands, and, as soon as the bipod touches the ground, shifts the left hand to the small of the butt, drops on his right hand, shooting out his legs to the rear, and gets into correct firing position;

(*c*) Raises sight, rotates magazine, pulls back cocking handle and lays on target indicated.

No. 2 (*a*) Picks up magazine carrier, runs forward in rear of No. 1 and lies down on left of gun; adjusts bipod (if necessary), and takes one magazine out of carrier.

(*b*) When No. 1 is ready to fire, holds out his hand and watches commander for signals.

When No. 1 and No. 2 get into position the remainder of the squad comes forward on the right of the gun to watch.

7. *To Fire.*—When the commander gives the signal " Fire," No. 2 touches No. 1 gently. No. 1 presses trigger and fires in bursts of about one second each, checking his aim after each burst, but allowing cocking handle to remain forward.

If fire is not opened within three seconds of the signal, the instructor must ascertain the reason for the delay.

8. *To Change Magazines.*—On the command " Change," No. 1 grips magazine with right hand, releasing catch with thumb. No. 2 helps to lift magazine off gun by pressing up centre block with left hand, and puts full magazine on gun with right, pressing it down carefully. No. 1 passes empty magazine (rim pointing upwards) *under* gun to No. 2, who replaces it in carrier; rotates new magazine and pulls back cocking handle. He then relays and continues firing.

9. *To Cease Fire.*—On the signal " Cease fire," No. 1 pulls back cocking handle and raises safety catch. Nos. 1 and 2 change magazine.

Before dropping safety catch again, No. 1 must pull back the cocking handle, to make sure that the sear is engaged.

10. *Out of Action.*—On the signal " Out of action "—

No. 1 (*a*) Unloads, *i.e.*, removes magazine, pulls back cocking handle, takes aim and presses trigger. Then lowers leaf of backsight.

(*b*) Retires with gun to original position, placing gun on ground.

No. 2 (*a*) Helps No. 1 to unload, and replaces magazine in carrier;

(*b*) Retires with No. 1, and places magazine carrier and spare parts bag two paces on left of gun. All take up original positions.

11. *To Change Round.*—On the command " Change round," No. 1 turns to the right and doubles round behind the squad and takes his place on the left of the line. The remainder step one pace to the right. The whole squad then re-numbers. Drill then continues as before, each man becoming No. 1 and No. 2 in turn.

APPENDIX IV.

INSTRUCTIONS FOR WEARING THE EQUIPMENT FOR CARRYING LEWIS GUN MAGAZINES. (S.S. 194.)

Four pouches constitute the set.

Each pouch can be used to hold either one or two magazines, according to the distance to be traversed.

The pouches are suspended in pairs over each shoulder, and should be connected to each other by the adjustable web straps attached to each pouch, in such a manner as to form a belt round the body. The web strap over the shoulder can be kept in position by the shoulder strap of the coat.

The essence of this equipment is that it leaves the hands free to use the rifle, and can be so adjusted by means of the web connecting straps as to clear the S.A.A. pouches and allow of ammunition being extracted from them without taking off the carrier.

AMENDMENT TO

S.S. 197—" THE TACTICAL EMPLOYMENT OF LEWIS GUNS."

Section IV.B. " In Open Warfare."

Delete para. 4 and *substitute:*—

4. In open fighting both sides will strive to obtain superiority of fire as early as possible; the mobility and concentrated fire power of Lewis guns will materially assist the remainder of the infantry in this task.

In trench warfare it may be inadvisable to send forward Lewis guns with the leading line, owing to the short distance to be traversed and to the fact that there are obstacles such as barbed wire to be passed. In open warfare, however, where attacks will probably be made over long distances and where the enemy's position may be only hastily entrenched, it will be possible to push forward Lewis guns with the leading line, the gunners moving and appearing to the enemy as ordinary riflemen. Lewis guns pushed forward in this way will facilitate the advance of the infantry by their covering fire.

In open fighting it will frequently be possible for Lewis gunners to precede the attacking infantry and to begin the establishment of a fire superiority under cover of which the infantry can advance : the enemy has made constant use of his light machine guns in this manner to establish " offensive points " in advance of his infantry, and opportunities to employ similar tactics against him should be carefully looked for and seized whenever the ground is suitable.

Opportunities for pushing forward Lewis guns may also occur when an advance is checked or a hostile counter-attack develops. In this way, superiority of fire may often be obtained without incurring the loss incidental to sending out strong reinforcements to the front line.

The formations of the leading infantry will vary according to the ground and the tactical situation, and it is not possible to lay down definite rules to suit all circumstances. The guiding principle, however, is that Lewis guns must be used in that position in which they can best assist the advance of the remainder of the infantry : this may be either in advance of, or from the flank or rear of, or actually with the attacking infantry, according to the configuration of the ground and the nature of the opposition.

PRINTED IN FRANCE BY A.P. AND S.S.   PRESS A—5/18—6341S—75,000